The INSIDE GUIDE

SPACE SCIENCE

Comets and Asteroids

By Jennifer Lombardo

Cavendish Square

Published in 2024 by Cavendish Square Publishing, LLC
2544 Clinton Street Buffalo, NY 14224

First Edition

Website: cavendishsq.com

Library of Congress Cataloging-in-Publication Data

Names: Lombardo, Jennifer, author.
Title: Comets and asteroids / Jennifer Lombardo.
Description: Buffalo, NY : Cavendish Square Publishing, [2024] | Series: The inside guide: space science | Includes bibliographical references and index.
Identifiers: LCCN 2023024087 | ISBN 9781502670106 (library binding) | ISBN 9781502670090 (paperback) | ISBN 9781502670113 (ebook)
Subjects: LCSH: Comets–Juvenile literature. | Asteroids–Juvenile literature. | Meteors–Juvenile literature.
Classification: LCC QB721.5 .L66 2024 | DDC 523–dc23/eng/20230815
LC record available at https://lccn.loc.gov/2023024087

Editor: Jennifer Lombardo
Copyeditor: Jill Keppeler
Designer: Deanna Lepovich

The photographs in this book are used by permission and through the courtesy of: Cover, pp. 27, 29 (bottom) Dima Zel/Shutterstock.com; p. 4 AAQ/Shutterstock.com; p. 6 (top) Pike-28/Shutterstock.com; p. 6 (bottom) jorisvo/Shutterstock.com; p. 7 Jim Cumming/Shutterstock.com; p. 9 shooarts/Shutterstock.com; p. 10 NASA/Goddard/University of Arizona; p. 12 Vector Tradition/Shutterstock.com; p. 13 NASA/ZTF/Caltech Optical Observatories; p. 14 Mirko Graul/Shutterstock.com; p. 16 JLStock/Shutterstock.com; p. 17 Stig Alenas/Shutterstock.com; p. 19 Matt Makes Photos/Shutterstock.com; p. 20 Makarov Konstantin/Shutterstock.com; p. 21 OskarWells/Shutterstock.com; p. 22 Peter Hermes Furian/Shutterstock.com; p. 23 Framalicious/Shutterstock.com; p. 24 Daniel Eskridge/Shutterstock.com; p. 25 (main) Jeff Caverly/Shutterstock.com; p. 25 (inset) Danny Ye/Shutterstock.com; p. 28 (top) Brian Donovan/Shutterstock.com; p. 28 (bottom) Nazarii_Neshcherenskyi/Shutterstock.com; p. 29 (top) Gerald Peplow/Shutterstock.com.

CPSIA compliance information: Batch #CWCSQ24: For further information contact Cavendish Square Publishing LLC at 1-877-980-4450.

Printed in the United States of America

CONTENTS

Early astronomers used instruments such as this astrolabe to figure out the position of the stars and planets. Today, thanks to telescopes and other technology, people know much more about space.

Chapter One

WHAT IS A COMET?

For thousands of years, humans have looked at the stars. Early peoples told stories about what they saw there, and the first astronomers began recording their observations around 1000 BCE. However, they didn't always understand what they were seeing. For example, although humans have seen comets in the sky since our earliest days, it took a lot of time and effort for people to figure out what they were.

Comets Throughout History

Early peoples such as the ancient Egyptians and Romans studied the sky in detail. They knew its changes throughout the year, and they named many of the stars, planets, and **constellations** after figures from their myths. However, unlike the rest of the bodies in the sky, comets showed up at **unpredictable** times. For this reason, many people thought they were bad **omens** or messages from the gods.

> **Fast Fact**
>
> The first known astronomical records came from the civilization of the Assyro-Babylonians. They lived in Mesopotamia, in what is now the country of Iraq.

Constellations such as Taurus (*shown here*) are generally in a fixed location, so astronomers know where to find them in the sky.

It wasn't until 1705 that people started to better understand comets. That year, English astronomer Edmond Halley used scientific **theories** created by Isaac Newton to figure out that some comets, which were later called periodic comets, passed Earth more than once. Before this discovery, people believed that comets never returned. Halley

Fast Fact

The comet Halley predicted is officially known as 1P/Halley. The "P" stands for "periodic," and the 1 shows that it was the first periodic comet ever discovered. People often call it "Halley's comet."

The Bayeux tapestry shows Halley's comet in the sky about 1066 CE—long before it was given its name.

believed comets that had been seen in 1531, 1607, and 1682 weren't three different comets, as people had thought—they were all the same comet, which came back around three times. He used his calculations to successfully predict that this comet would come back again in 1758. Because of his discovery, the comet was named after him.

A Comet's Makeup

Comets are made of leftover material from the creation of the solar system 4.6 billion years ago. They are often called "dirty snowballs" because they're made of ice and a little bit of rock. Each comet has a center called the nucleus, which is made of frozen gas and dust.

Fast Fact

Many scientists believe part of the reason life arose on Earth was because comets crashed into the young planet and deposited, or left behind, water and other important compounds.

Many people gathered to watch Comet NEOWISE as it passed overhead in 2020. It won't return for another 6,800 years.

HOW DO WE KNOW?

If comets are so far away and move so fast, how do we know what they're made of? Scientists had to launch spacecraft to find out! In 1986, astronomers from many different countries worked together to take photos of Halley's comet's nucleus. In 2004, a NASA spacecraft called *Stardust* flew within 147 miles (236 kilometers) of Comet Wild 2 and returned to Earth in 2006 with samples for the scientists to study.

Another mission sent the spacecraft *Deep Impact* crashing into a comet called Tempel 1, taking photos and samples as it did so and sending them back to Earth. In 2011, NASA returned to Tempel 1 to see how it had changed over the years.

When a comet passes close to the sun, the sun's heat warms the nucleus, changing the ice back into a gas. This creates a layer around the nucleus called a coma. As the comet continues to move through space, its coma may be blown backward, giving some comets the long, bright tail people see when they pass over Earth.

Where Do Comets Come From?

Beyond Neptune is an area of space known as the Kuiper Belt. Pluto and Eris, two **dwarf planets**, are the largest objects in the Kuiper Belt. Many of the other objects are balls of ice and rock. When gravity pulls these icy objects into orbit around the sun, they become short-period comets.

This picture shows where the Kuiper Belt and Oort cloud are located.

Because the Kuiper Belt is relatively close to the sun, it takes short-period comets less than 200 years to make a complete orbit. Long-period comets are thought to come from a more distant part of space known as the Oort cloud. They're so far away that some can take up to 30 million years to orbit the sun just once!

This picture, taken by NASA in December 2018, shows four views of an asteroid called Bennu.

Chapter Two

WHAT IS AN ASTEROID?

Like comets, asteroids are leftovers from the beginning of the solar system. They orbit the sun. However, while comets are mostly ice, asteroids are mostly rock and metal. Asteroids are much harder to see than most comets because they aren't as reflective.

Pulled Out of Orbit

Asteroids come from an area of space known as the asteroid belt, which is located between Mars and Jupiter. Like everything else in our solar system, the asteroid belt orbits the sun. One reason why asteroids have less consistent orbits than comets is that their orbits can be changed by Jupiter's gravity. When an asteroid in the belt passes Jupiter in its orbit, Jupiter's gravity can pull it away from the belt and send it flying across space in a new orbit.

Fast Fact

NASA keeps track of any asteroids it finds. As of 2023, there were more than 1.2 million known asteroids in our solar system.

Types of Asteroids

Asteroids can be grouped into different **categories**. One category is for where the asteroid orbits. Most are in the main

This picture shows where the asteroid belt is in our solar system.

asteroid belt. Some, called Trojans, share a planet's orbit around the sun. Most Trojans share Jupiter's orbit, but some orbit with Mars or Neptune. In 2011, scientists found a Trojan orbiting the sun in the same path as Earth.

Fast Fact

Scientists believe there are as many Jupiter Trojans as there are main belt asteroids.

Trojans follow the same orbit as the planet they're with, so they can never hit the planet. However, not all asteroids near a planet are Trojans. Scientists keep an eye out for near-Earth objects, including asteroids and comets. Near-Earth asteroids are ones that pass close

This picture, taken through a telescope, shows a near-Earth asteroid. Early astronomers gave asteroids their name, which means "starlike," because they thought they looked like stars.

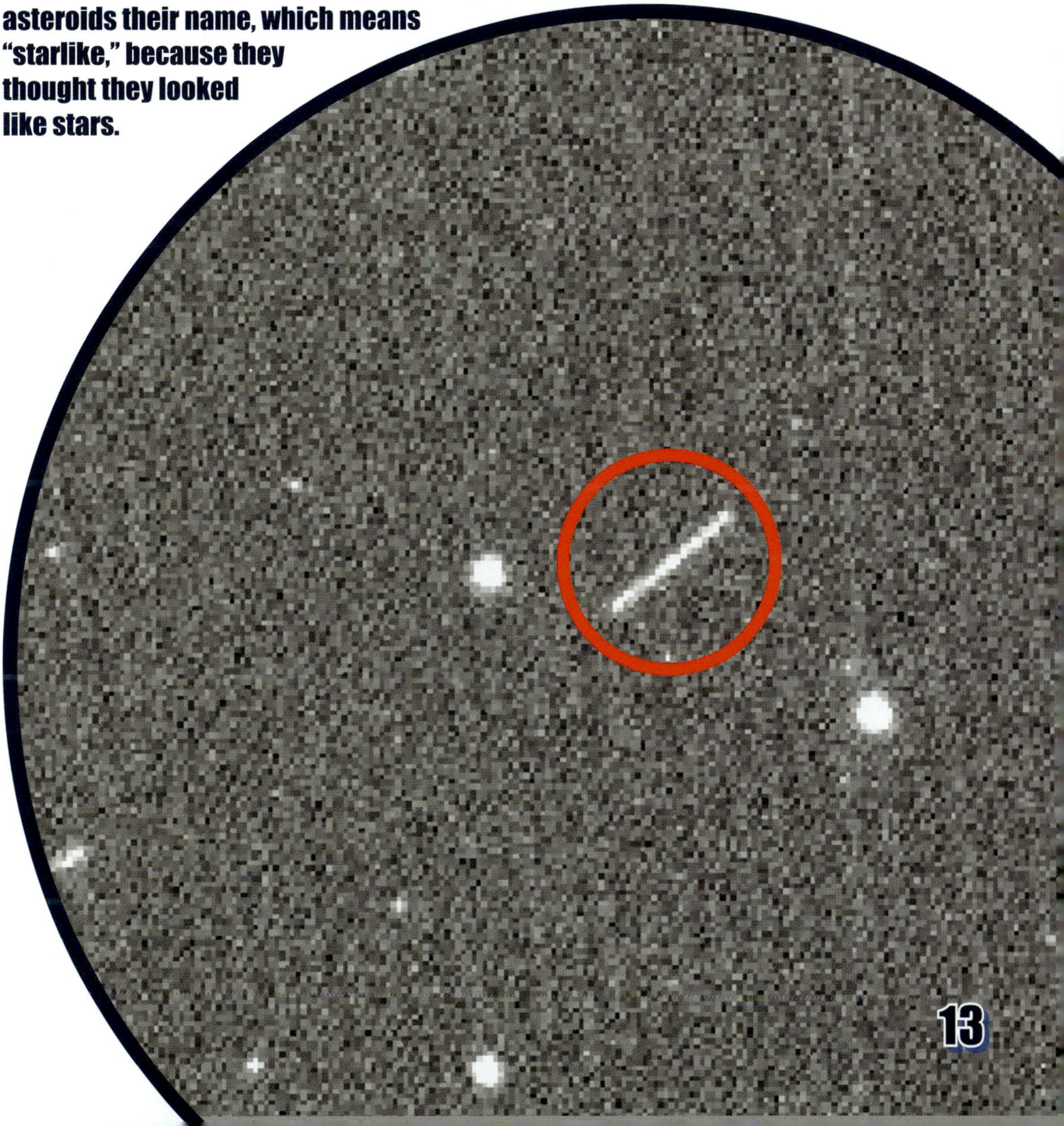

CHONDRITES AND CHONDRULES

A chondrite is a type of rock that only forms in space. It gets its name from the chondrules that are embedded, or stuck, in it. Chondrules are round and very tiny; even the largest ones can only be measured in millimeters. They formed when the rock and metal in an asteroid melted under intense heat and then cooled.

Scientists call chondrites the "building blocks" of the solar system. In other words, they're made of the same matter that came together to form the planets and moons. Most of the leftover pieces became comets and asteroids. Some scientists believe it was chondritic asteroids, not comets, that brought life-forming compounds to Earth.

This photo shows a chondrite. The circles in it are individual chondrules.

enough to the planet to possibly be drawn in by its gravity and hit it. Most harmlessly pass by millions of miles away.

Another way scientists categorize asteroids is by their makeup, which is determined by how far from the sun the asteroid formed. The most common type of asteroid is called C-type. The "C" stands for "chondrite." C-type asteroids are made up mainly of clay and minerals called silicates. They're some of the oldest objects in the solar system.

The second-most common asteroids are S-type. These are stony asteroids. They're made of silicates and a nickel-iron **alloy**. The least common, M-type, are metallic. They're made up entirely of nickel-iron. Asteroids can be one large piece of solid rock or "rubble piles," which are many smaller rocks held together by gravity.

Fast Fact

The largest known M-type asteroid is about 120 miles (193 km) wide. That's more than three times as wide as the state of Rhode Island!

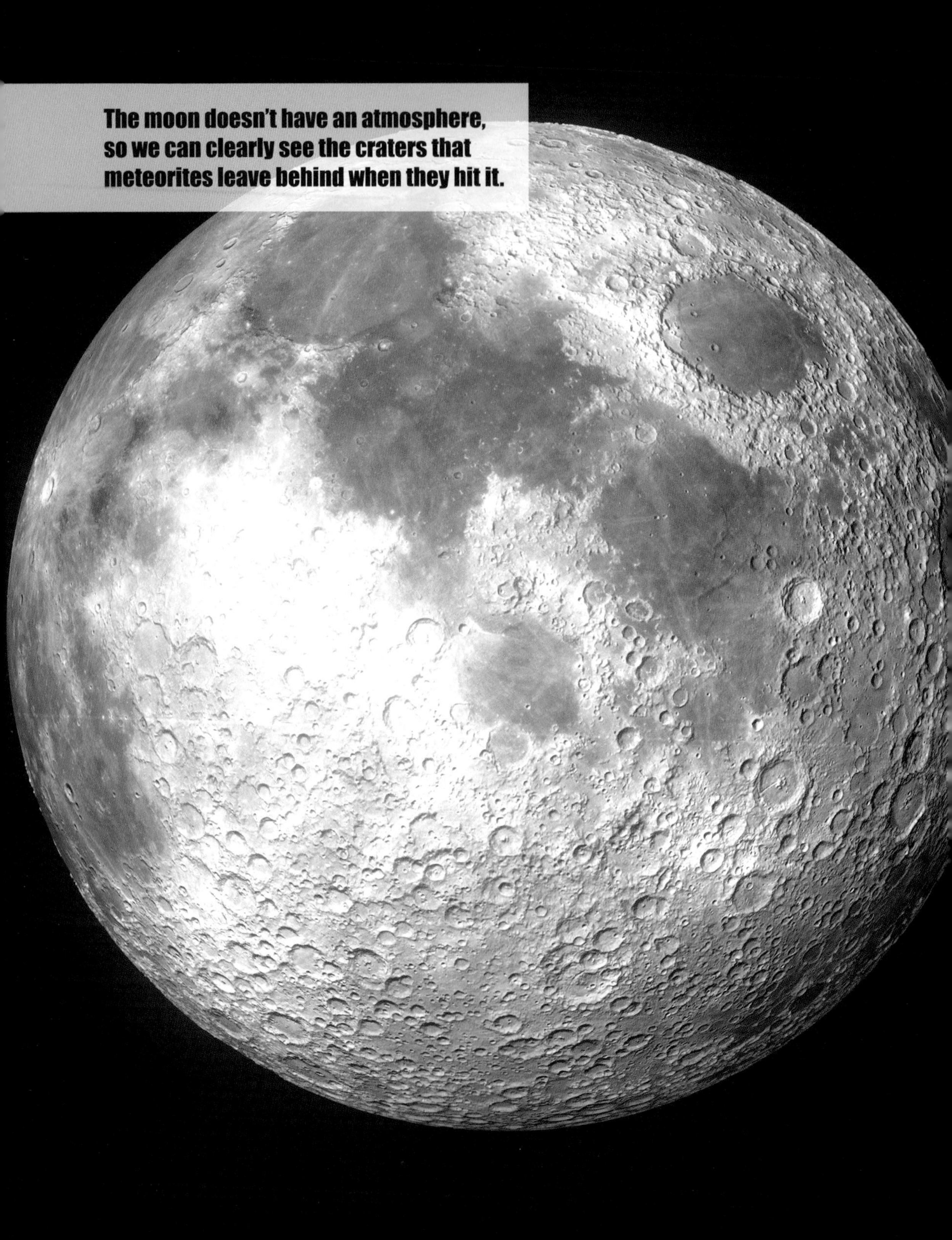

The moon doesn't have an atmosphere, so we can clearly see the craters that meteorites leave behind when they hit it.

METEORS AND MORE

Chapter Three

Believe it or not, Earth gets hit by comets and asteroids—or at least pieces of them—quite often! Most of these are so small that they burn up in the atmosphere, creating a streak of light across the sky. This is what people often call a shooting star, but scientists call it a meteor. When a meteor hits Earth, the only thing that falls to the ground is dust and other small particles that are left over after it burns.

Before a meteor enters the atmosphere, it's called a meteoroid. Meteoroids are pieces that break off asteroids and comets. If a

Fast Fact

Scientists first learned about chondrites and chondrules by studying chondritic meteorites.

This is a piece of a meteorite that landed in what is now Cape York, Greenland, up to 10,000 years ago. For many years, the local Inuit used the iron in it to make tools.

meteoroid is big enough, not all of it burns up in the atmosphere as a meteor. What's left is a rock that falls to the ground, when it becomes known as a meteorite.

Many Meteorites

Most people have never seen a meteorite hit the ground. This might make it seem like it doesn't happen very often. However, a team of British scientists found that Earth is hit by about 17,000 meteorites each year! The scientists spent two years in Antarctica searching for meteorites. They chose Antarctica because the meteorites were easier to see against the snow. After they counted how many meteorites they found in Antarctica during those two years, they were able to use math to make an estimate, or close guess, about how many fell all over the planet in that same time period.

One reason why people don't often find meteorites is because most of them are small. According to NASA, anything smaller than about 82 feet (25 meters) wide will burn up as a meteor, with nothing more than tiny rocks and dust reaching the ground. Most meteoroids are even smaller than this. Another reason is because Earth is huge! Most of the planet is covered in water, so that's where most meteorites

Fast Fact

In 2013, a meteor the size of a house exploded in the sky over Chelyabinsk, Russia. Even though it didn't hit the ground, the meteor caused damage to buildings and injuries to many people.

The white snow of Antarctica makes dark rocks such as meteorites stand out more clearly.

THE PERSEID METEOR SHOWER

In 1862, two astronomers named Lewis Swift and Horace Tuttle discovered a short-period comet with a nucleus about 16 miles (26 km) wide. The Swift-Tuttle comet takes 133 years to complete its orbit. The last time it passed by Earth was in 1992. However, every time the comet passes by, it leaves behind meteoroids.

In 1865, astronomer Giovanni Schiaparelli realized that the Perseid meteor shower, which happens every summer, comes when Earth passes through the part of the sky where these meteoroids float. This meteor shower gets its name because the meteors appear in the sky near the constellation Perseus. However, because comets aren't part of constellations, the meteors don't actually come from Perseus.

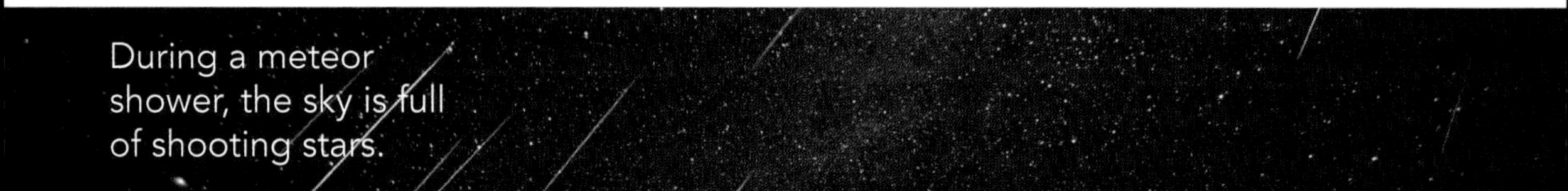

During a meteor shower, the sky is full of shooting stars.

Fast Fact

The largest meteorite ever found is known as the Hoba meteorite. It landed in Hoba West, Namibia, about 80,000 years ago. It's been dug up but never moved. It weighs about 132,000 pounds (59,800 kilograms)!

land. The ones that land on solid ground often blend in with the rocks that were already there, especially if the landing area is somewhere not many people live or visit often.

Areas such as parts of the Australian outback (*shown here*) aren't often visited by people. A small meteorite could go completely unnoticed if it fell in such a place.

This map shows where the Chicxulub **impactor** hit.

PREPARE FOR IMPACT!

Chapter Four

With so many comets, asteroids, and meteoroids whizzing through space and hitting our planet, scientists have realized that a large meteorite strike is unavoidable. In fact, it's already happened in the past. The most **infamous** meteorite is called the Chicxulub impactor. Scientists gave it this name because it landed near what is now the town of Chicxulub, Mexico.

Fast Fact

For many years, scientists couldn't agree on whether the Chicxulub impactor was a comet or an asteroid. By studying the makeup of the rocks at the impact site, they realized it was most likely an asteroid.

Large meteorites leave behind large craters, such as this one in Arizona. Over time, they become less noticeable. This happens more quickly in places that get a lot of rain.

The End of the Dinosaurs

The Chicxulub impactor landed about 66 million years ago. It was almost 9 miles (14.5 km) wide, and the crater it left behind—most of which is underwater—is about 93 miles (150 km) wide and 12 miles (19 km) deep. Scientists didn't find it until 1978 because it's so large that it can only fully be seen from the air. Additionally, as time passed, the ground changed, smoothing out and filling in the crater so it was less noticeable.

Fast Fact

According to NASA, any object larger than about 0.5 mile (1 km) wide that hits Earth could affect the entire planet. Main belt asteroids can be up to 583 miles (940 km) wide, but these are too far from Earth to be a danger to us.

The Chicxulub impactor changed Earth's climate by making it colder. This affected living things in major ways.

The only dinosaurs that survived the Chicxulub impactor were some of the beaked avian, or birdlike, dinosaurs. All birds on Earth today, from chickadees (*top*) to cassowaries (*right*), are the descendants of those dinosaurs.

The meteorite's impact sent liquid rock and steam high into the air. Smoke and dust filled the air, blocking sunlight and causing the planet to get colder over time. It also created a tsunami, or a series of waves, that flooded the world. Because of all these changes, scientists believe about 75 percent of the plant and animal life on Earth at the time—including most of the dinosaurs—died.

A RARE EVENT

On May 8, 2023, the Kop family was startled when a small rock—about 4 inches (10 centimeters) by 6 inches (15 cm)—crashed through their ceiling in New Jersey at 1 p.m. It hit the floor in one of the bedrooms, bounced to the ceiling, and then came to rest in a corner. The bedroom was damaged, but no one was hurt.

This small rock turned out to be a meteorite. Scientists believe it's between 4 and 5 billion years old and may have come from the Eta Aquarid meteor shower. This meteor shower happens every year between April 19 and May 29. The meteoroids Earth passes through at this time were left behind by Halley's comet.

Stopping the Next Impact

Because scientists know that another meteorite could hit at any time, they've spent years getting ready for it. According to NASA, a meteoroid big enough to cause serious problems passes Earth once about every 2,000 years. Luckily, none of them have hit the ground in recent times.

Astronomers all over the world keep track of near-Earth objects with telescopes. They track them and report when the danger is over. Comets are much easier to track than asteroids because many move in predictable orbits and reflect a lot of light.

Scientists have several different ideas about how to stop a large meteoroid before it hits the ground. One idea is to blow it up in space! The idea is that a bomb could break it into smaller chunks that would burn up harmlessly as meteors. However, it's possible that the chunks wouldn't be small enough to keep from causing damage.

Another idea is to heat a meteoroid with lasers. This might burn off its outer layers and make it smaller. A third idea is to hit it with an unmanned spacecraft to change its direction and send it away from Earth. Scientists are always working on new ideas and creating new technology to keep our planet safe from comets and asteroids.

Fast Fact

Scientists believe an asteroid would have to be about 60 miles (96 km) wide to kill every living thing on the planet. Luckily, there are no near-Earth objects anywhere near this size!

Deep-space telescopes such as the James Webb Telescope can help scientists spot comets, asteroids, and meteoroids before they pose a danger to Earth.

THINK ABOUT IT!

1. How do you think Edmond Halley figured out which years the comet named for him had passed over Earth in the past?

2. Why don't asteroids have a tail like many comets?

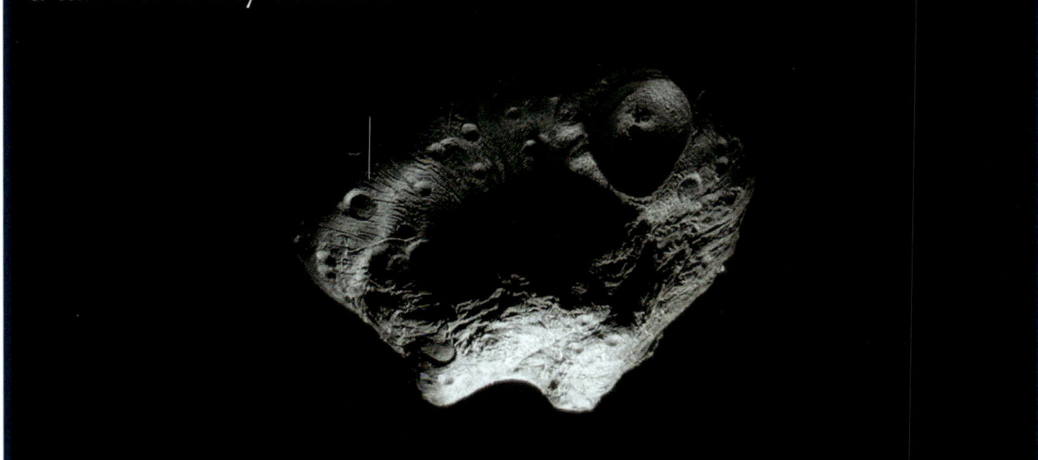

3. Why are asteroids more likely to become meteorites than comets are?

4. Can you think of other ways a meteoroid could be stopped from hitting Earth?

GLOSSARY

alloy: A mixture of two or more metals.

category: A way of grouping similar things.

constellation: A group of stars that forms a pattern.

dwarf planet: A celestial object similar to a planet but too small to clear other objects from its orbit.

impactor: Something that hits something else.

infamous: Well known in a bad way.

omen: A sign of things to come.

tapestry: A heavy cloth that has designs or pictures woven into it and is used as a wall hanging.

technology: The use of science to solve problems, or the tools used to solve those problems.

theory: A general principle or set of principles that explains facts or events of the natural world.

unpredictable: Describing someone or something whose actions are difficult or impossible to guess ahead of time.

FIND OUT MORE

Books

Kurtz, Kevin. *Comets and Asteroids in Action: An Augmented Reality Experience*. Minneapolis, MN: Lerner Publications, 2020.

Lawrence, Ellen. *Comets, Meteors, and Asteroids: Voyagers of the Solar System*. Chicago, IL: Sequoia Kids Media, 2022.

Watson, Fred. *Spacewarp: Colliding Comets and Other Cosmic Catastrophes*. Sydney, Australia: NewSouth Publishing, 2021.

Websites

BrainPOP: Comets
www.brainpop.com/science/space/comets
Watch a video, play games, and take a quiz to test your knowledge of comets.

Ducksters: Asteroids
www.ducksters.com/science/physics/asteroids.php
Read more cool facts about asteroids.

NASA Space Place: Comets
spaceplace.nasa.gov/menu/comets
Learn more about comets, asteroids, and meteors.

INDEX